I0896349

A GUIDE TO METAL DETECTING

A GUIDE TO METAL DETECTING

GRAEME RUSHTON

WHITE OWL

AN IMPRINT OF PEN & SWORD BOOKS LTD.
YORKSHIRE – PHILADELPHIA

First published in Great Britain in 2021 by
Pen & Sword WHITE OWL
An imprint of
Pen & Sword Books Ltd
Yorkshire – Philadelphia

ISBN 9781526780683

A CIP catalogue record for this book is available from the British Library.

Printed and bound in India by Replika Press Pvt. Ltd.

Pen & Sword Books Limited incorporates the imprints of Atlas, Archaeology, Aviation, Discovery, Family History, Fiction, History, Maritime, Military, Military Classics, Politics, Select, Transport, True Crime, Air World, Frontline Publishing, Leo Cooper, Remember When, Seaforth Publishing, The Praetorian Press, Wharncliffe Local History, Wharncliffe Transport, Wharncliffe True Crime and White Owl.

For a complete list of Pen & Sword titles please contact:
PEN & SWORD BOOKS LIMITED
47 Church Street, Barnsley, South Yorkshire, S70 2AS, England
E-mail: enquiries@pen-and-sword.co.uk
Website: www.pen-and-sword.co.uk

Or
PEN AND SWORD BOOKS
1950 Lawrence Rd, Havertown, PA 19083, USA
E-mail: Uspen-and-sword@casematepublishers.com
Website: www.penandswordbooks.com

FSC
www.fsc.org
MIX
Paper from responsible sources
FSC® C016779

Contents

Foreword

THIS BOOK IS PRIMARILY written for anyone who is thinking about starting the wonderful hobby of metal detecting, or for people who have had some experience and would like to learn a little more. It will also be of interest to those who would like to find out more detailed information about this fascinating hobby.

The book aims to take individuals through all the stages of metal detecting, with each chapter carefully giving advice and guidance on how to get the best out of the hobby: from purchasing your own equipment to gaining your first permission to detect.

Throughout the book there will be many handy hints and tips. It is hoped that the reader quickly gains the knowledge and knowhow to be successful and at the same time enjoy the experience.

The author is a seasoned detectorist with over thirty-nine years experience; he regularly holds classroom talks with school children, showing the historic finds he makes and also information on the history behind them. He travels all over the country and may often be seen out in the fields using his detector to recover ancient coins and artefacts.

Introduction

METAL DETECTING AS a hobby has been around for many decades, in fact the very first detector was made in 1881 by Alexander Graham Bell, who invented a rather crude device to locate a bullet that was embedded in President James Garfield! This was a far cry from the modern, highly sophisticated machines that we see today. Metal detectors are manufactured in several different countries, including France, the United States, Turkey, Australia and the United Kingdom, and the hobby is enjoyed by many across the globe.

It is of course the fascination of finding buried coins and artefacts from the past that drives people to spend many hours out in the fields, woodlands and beaches in the hope of unearthing something that our ancestors once held; from the medieval silver penny to the Victorian brooch, the finds that come to light are varied and the age range vast.

In the past, a detectorist was associated with being a bit of a 'Geek', someone who probably also enjoyed train-spotting and other semi-eccentric hobbies. However, over the last twenty years or so more and more people have taken up the pastime. This is partially due to the fact that there have been some incredible discoveries such as the Staffordshire Hoard and the Hoxne Hoard from East Anglia, which hit the media and captured people's imagination almost instantly. We now see families, every weekend trailing up and down fields in the hope of finding a coin, artefact or, dare we dream, a significant hoard!

Of course most of us who use a metal detector never find the 'Big One' and only a handful of individuals really make the headlines, but unearthing a Roman coin or a medieval buckle captures our imagination of the distant past in such a way that we will be out every week, in all weathers doing our best to discover and preserve history.

Metal detecting plays a vital role in bringing important discoveries to light; without metal detectorists pursuing their hobby week in week out, some of the amazing historical artefacts and hoards now on display in museums up and down the country would never have been brought to light.

So how do you take your first footsteps into this amazing hobby? You will no doubt be wondering where to start. Is it a simple matter of wandering down the beach on a sunny day with the detector in hand or a search of the local farmer's fields at the weekend, or on your day off? Well there is a little bit more to it than that; hopefully this book will go some way to help answer all your questions!

The author enjoying an evening's detecting.

CHAPTER ONE

Choosing the Right Detector & Equipment: Picking a Suitable Machine

THIS CHAPTER WAS always going to be a tricky one, but let's make it as simple to digest as possible. Going back a few decades or so, the choice for the beginner was a relatively simple one, indeed when I took my first tentative steps into the hobby there were only really a handful of metal detectors to choose from. These arrived plainly boxed from the manufacturer with a paper pamphlet inside giving the user a brief overview on how to use the detector and that was pretty much that: no decent books; no online forums to ask questions; no user groups; no YouTube channels showcasing that particular brand; it was simply a matter of using it as often as possible and getting used to the metal detector's features and characteristics; very much like riding a bike, the more you did it, the better you became!

Fast forward in time and we now see an eye-watering selection of metal detectors on the market. These are manufactured in countries such as America, Great Britain, France and Australia, with the makers producing many varied models from introductory brands to professional units, some are simple to use, others, it seems need an engineering degree to master!

So let's break down the common myths that exist. Firstly, in my opinion you don't necessarily need to break the bank to own a metal detector. Indeed, some of the introductory units on the market currently have similar performance levels to machines twice their price, maybe slightly fewer features than the more expensive models, but let's face it, there is nothing more off-putting than

Venue chosen, ready to go!

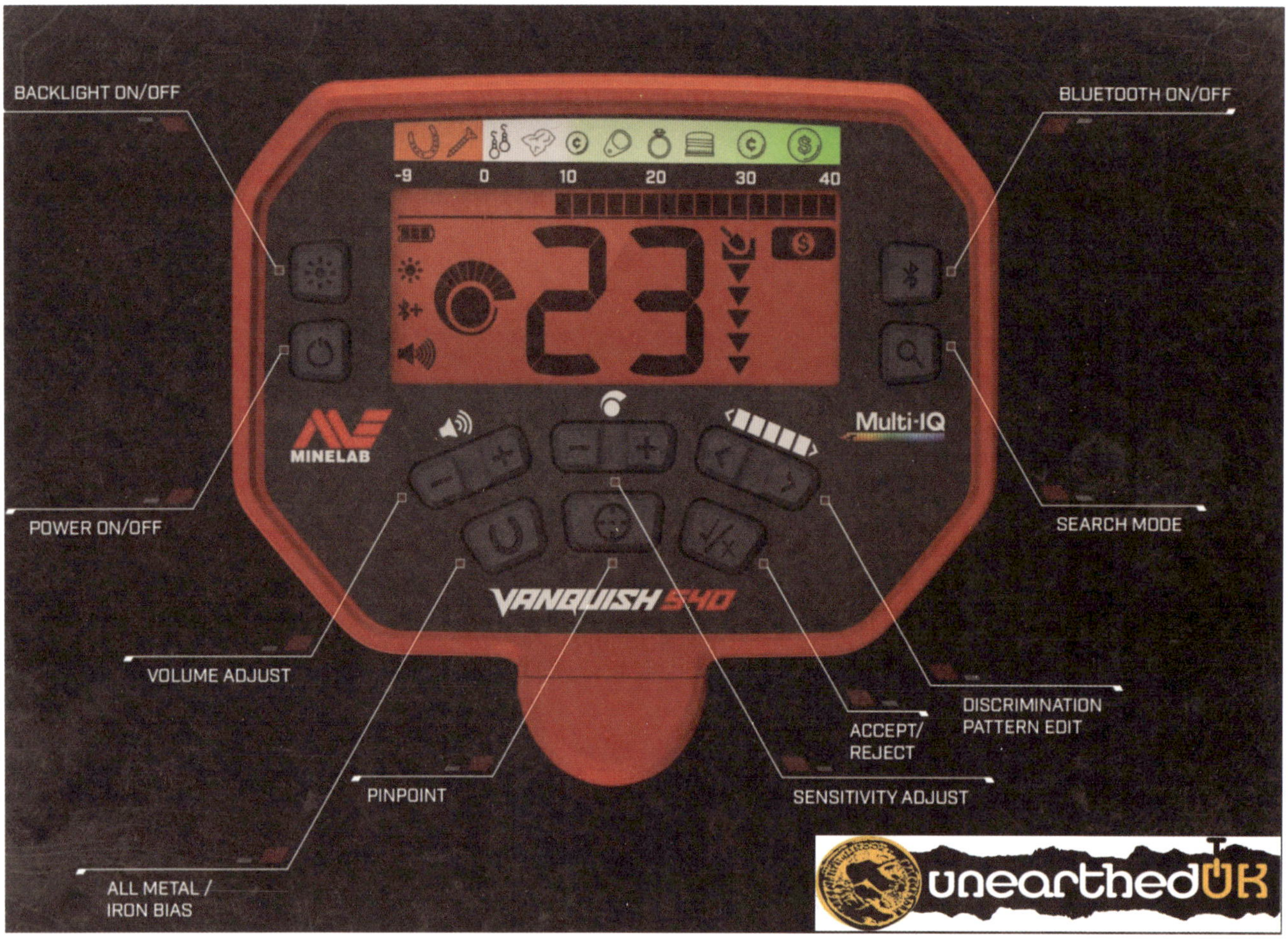

Control box of a metal detector showing its features.

trying to use a product that makes you scratch your head more than actually using it. The key to being successful in the hobby and with your new purchase is quite simply this:

SIMPLICITY

No need to over-complicate things with added features on the detector that you will seldom use. To be able to adjust and set the machine quickly and easily is always the best option.

LIGHTWEIGHT

The product has to be light enough for you to swing out in the fields without suffering from fatigue; and be nicely balanced so you can actually enjoy the experience of walking tirelessly up and down a ploughed field or on a windswept beach for hours on end. So make it a comfortable one. Nowadays there is no excuse for manufacturers to make heavy, cumbersome machines.

PERFORMANCE

You can have the most feature-packed detector on the planet, but without the performance to match it then it's a pretty worthless purchase. A detector has to be able to deal with coins and artefacts buried in and around iron, so the recovery speed must be decent. It must have reasonable depth capabilities to pick up objects buried under pasture fields, and – let's not forget – if you are wanting to go down the route of being a beach detectorist then your machine must be able to cope with the high levels of mineralisation from the salt in the wet sand. We can cover some of the technical features of detectors later in this chapter

COST

The detector must tick the boxes mentioned above and also come within your budget. As I touched on earlier, you don't need to break the bank to purchase a decent metal detector; a mid-range machine (which is what my own personal detectors fall into) will be more than capable of finding the coins and artefacts that a lot of the high-end units do.

So what do you need to look for specifically? Feature-wise what do you need in a detector?

RECOVERY SPEED

This is very important, especially in iron-infested areas for example. In a nutshell and without getting too technical, this feature determines how quickly a detector can register several targets/items buried close together, so it's wise to look for detectors that the manufacturer claim to have this as one of their key features.

SENSITIVITY

This function is important for a number of reasons, including depth and the detector's ability to find small targets for example. Remember the higher you go sensitivity-wise the deeper the metal detector can find potential coins and artefacts; in other words, when you increase this function you are effectively asking the metal detector to hit those weak signals that could well turn out to be that desirable coin you are searching for. However, the more you crank it up, the more unstable the machine can become. Finding a happy medium is the key.

DISCRIMINATION

This function allows you to hear certain responses from targets that you want, and reject others such as iron. It may take you some time to find the ideal setting: for example if you set the discrimination too high you risk rejecting desirable metals such as gold, however, too little 'discrim' can see you digging up a whole host of undesirable finds, including iron nails, horse shoes etc.

VOLUME

Not all metal detectors have this function but it is important to touch on it. If your detector does have this function it allows you to control the volume of target sounds, the higher you set it, the louder the noise made when it discovers a potential buried coin or artefact. Some volume features on metal detectors also control iron volume, from a subtle noise in the background to a loud, slightly annoying noise! Again personal preference takes precedence here.

GROUND BALANCE

This function allows the detector to harmonize with the varying ground conditions and eliminate any spurious signals.

PINPOINT FUNCTION

This is a feature that most modern metal detectors have; quite simply it is used to determine the exact location of a buried target, usually by the detectorist holding down the pinpoint button and sweeping over the already located target. The tone-pitch of the detector will rise when you are directly over the target; tuning your ears into this sound is a relatively simple process to learn. This feature is widely used by most detectorists, however, some are skilled enough to use their eyes to line up the coil and the target without even touching the mode. Spending a little time with a coin on the floor and slowly using the pinpoint facility over it will help you understand the precise function of this feature.

MULTI-FREQUENCY & SINGLE FREQUENCY

It can be a wise move to check out if the metal detector you purchase is either multi- or single- frequency. You may be thinking what is that? Without getting too technical, the common definition of frequency is the number of electronic waves sent into the ground

A Norman Gargoyle Mount unearthed for one lucky detectorist.

to detect metal. To explain a little more about single frequency for example, one of the detectors I use is a single frequency 19 kHz model; being set at 19 kHz means the detector will send and receive 19,000 times per second, meaning it is very sensitive to small objects and gold – hence why I use it!

I find it accurate, easy to use and good at registering unwanted targets like iron. Just recently I have added a multi-frequency detector to my armoury, the reason being straightforward enough: this particular platform makes use of more than one frequency at a time making it accurate, deep and stable, and really good at target identification, and offering an added advantage of working very well on beaches and wet sand too. The modern multi-frequency detectors are something to behold once mastered and many desirable finds fall to these models each year.

TONES

Most modern metal detectors have a tone feature; for example metal objects such as iron nails will usually have a low tone or grunt, while more desirable objects such as gold and silver coins will give a higher-pitched tone. We class this as 'two-tone', but a growing number of machines now have multi-tone systems; indeed one of my earlier detectors, now a decade old, had a four-tone system which enabled me to distinguish from the low iron grunts, the mid-tones for say copper and bronze coins and then high tones, almost squeaks, for silver. This four-tone feature took a little time to get the hang of, but I was soon able to establish if I was going to be digging a silver hammered coin or a silver sixpence for example. Of course this wasn't always the case, but it was really accurate on the targets.

Other features that metal detectors may have are: a notch facility, all metal mode, terrain modes, battery indicator, backlight and tone break (including v-break systems), but without wanting to complicate this section too much, I think we have covered the main areas that will be important to you as a new user.

Detecting equipment is next to discuss and I will keep this straightforward. Most, if not all, beginners coming into the hobby will already have accessories and clothing at their disposal that they can use when venturing out to either the beach or local fields, so let's have a quick look at what is needed:

A strong spade is always advisable.

A SPADE

It is crucial to equip yourself with a decent, robust spade. Let's face it, you will not be able to unearth your finds without one! Over the years I have used many different types, including custom-made stainless steel ones that will last a lifetime, and small, lightweight shovel types that are easy to handle, carry and travel with. However, what some people fail to see is they already have a decent spade, probably hiding in their shed that will be used occasionally while working in the garden. As long as it is robust enough to dig neat holes and isn't too heavy, then you can easily match it across to metal detecting; the head of the spade must be sufficiently angled and not too big, so that the holes you dig are neat enough to ensure the landowner/farmer/permission holder is happy that you have made a tidy job of his/her land and fields.

CLOTHING

If you are interested in starting metal detecting then it is likely you are already the outdoor type, which means you will probably have a decent, weather-resistant jacket to protect you from the ever-changing climate. It is vital to keep yourself warm and protected; there is nothing worse than organising a day's detecting only to find the weather and the clothing you have letting you down. You want to have a positive experience rather than something that could well put you off when you get soaked through to the skin. Next up will be a decent pair of weatherproof, warm trousers, comfortable enough to walk in for a few hours at least and also be able to offer some protection from the elements. If possible look for ones that have built-in knee pads as these will kill two birds with one stone – bending down and being on your knees several times a day can soon take its toll – so the knee pads will offer welcome protection that your knees and joints will need. Footwear next, which of course is probably the most important part of your attire. Frustratingly I often see people out detecting with unsuitable footwear: training shoes or cold wellies that leak, or that are years old. These can completely ruin your day, as once your feet get wet and cold the rest of your body will follow suit. Choose yourself a decent pair of boots, thermal wellies or something of that nature that will firstly stand up to the weather, be tough enough to deal with the mud, sand and wet grass, most importantly offer the thermal warmth that will keep your feet nice and snug.

Boots to keep the feet dry and warm!

Another handy tip here is make sure you inspect the soles of the wellies or boots; remember you will be digging holes while detecting, and your left (or right) foot will be subjected to the weight of your body pressing down on your digging tool/spade, so the business end of the footwear must be strong enough to cope and not split. There is nothing worse than a small, almost invisible slit that will let moisture and cold seep in and ruin the whole experience.

Gloves are an essential accessory.

GLOVES

Once again a very important accessory to have, not only to keep your hands warm, dry and dirt free but also to give you the added protection against some of the nasty bacteria and bugs that can exist in the soil. Something else to consider and look out for are sharp objects – gloves will give you a barrier against glass, pottery, natural flint and stone and other potentially dangerous objects that can cut and hurt. Remember, a cut hand can ruin a day's detecting!

HEADWEAR

A cap in the summer to shade you from the sun and a woollen-type hat in the winter to keep the head warm must be part of your kit bag; most people have at least one of these items at home that can be used for a day's detecting.

HEADPHONES

Most successful and considerate metal detectorists use headphones with their machines. There are a number of important reasons why. Firstly imagine walking around in the peaceful countryside with your detector bleeping and sounding off every few seconds or so. This would not only disturb the people around you but could also

Detecting can be a social activity!

upset the landowner, plus there is nothing worse than detecting around an individual who doesn't wear them as it can ruin your own concentration levels and disturb the peace.

Another important reason to wear headphones is they enable you to hear the deeper targets that are buried, or smaller items that could be lying on edge in the soil; different tones will also be distinguishable, which could all lead to better finds. Using the detector and just relying on its built-in speaker instead will be of detriment to the user. Another important reason to wear headphones is battery life. Plugging in a set to the control box will save energy as no power is needed to drive the external speaker – so you are actually saving money.

Lastly but still equally as important, you are cancelling out any exterior noise. Imagine metal detecting alongside a busy road for example – the chances of hearing anything from your detector would be slim – but by plugging in a set of headphones you will be able to ignore the traffic and hear everything under your feet. The same applies to detecting on the beach, only this time it's the wind and wave noise you're blocking out.

PINPOINT PROBE

While some see this accessory as an expensive luxury, it makes sense to consider using one of these hand-held devices: quite simply a pinpoint probe or pinpointer will make the job of finding an object or a coin much easier, which in turn will save a lot of unnecessary digging. You simply switch on the device, and point it in the hole you have dug and by buzzing, beeping or vibrating it will take you directly to the potential find. Some have different features, most are waterproof and can be used in rivers and on beaches, all are fairly straightforward to use and there are many alternative models to choose from. It is worth spending the time looking at their various functions and choosing one that best suits your budget. A tip from me would be attach the device to a lanyard; that will save you from potentially losing the probe – believe me it's amazing how many people lose theirs out in the fields while detecting.

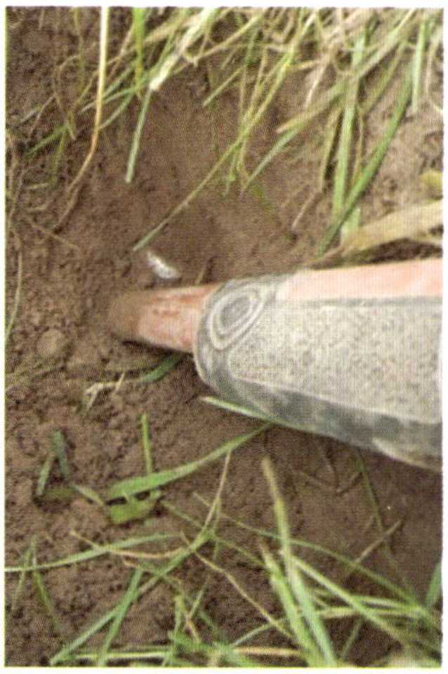

A pinpoint probe locating a target.

SAND SCOOP

If beach detecting appeals to you then it is worth considering purchasing a sand scoop. These accessories will help you recover your targets quickly and easily compared to a standard spade. They come in different sizes and shapes for specific jobs – from short-handled scoops, to medium ones, to long-handled versions. For example the long-handled scoops are very useful if you decide to search and wade in shallow water, where the medium-handled ones are easier to manipulate and are ideal for searching in the sand and shingle. All scoops are designed with filters or screens at the business end, this is to allow the small pebbles and sand to flow through while at the same time capturing coins, rings and other objects within the head.

Consider the stainless steel versions, they will give you a much longer shelf life than say mild steel versions; the salt water can and will damage the non-stainless examples very quickly, so it may well be false economy choosing a cheaper version. Some individuals go one stage further and design and make their own, detailing these accessories to fit their needs perfectly.

In summary, make sure you choose one that will suit the area of the beach that you wish to search. It is pointless to select a short-handled sand scoop if you wish to do some shallow water searching and the same goes for the long-handled one if you are scanning the dry sands.

BATTERIES

Every metal detector needs a battery to power it; some ultra-modern detectors have built-in ones, where you charge the unit very much like a mobile phone or other electrical devices. Most detectors can take rechargeable and disposable examples; either way choosing the right brand can sometimes be an advantage. I recall one gentleman using his brand new machine straight out of the box, all wired up and ready to go. However, things went downhill quickly when his new detector kept on shutting down after a few minutes. After helping him troubleshoot potential issues, we nailed it down to the fact he was using a cheap 9-volt battery from his local store. That was rapidly discarded and replaced by a much better brand; soon all was well with him and his detector.

He later told me over the phone that he bought three for £1, an offer which sounded too good to be true and ended up just that. It is worth paying the extra couple of pounds if you can to avoid issues such as this.

Rechargeable batteries are a pretty safe bet and you can purchase really good charging systems with the batteries added. This way you can charge them in your own time and it will save you money long term, however, j, be aware that rechargeable batteries can drop off power-wise rather sharply, so it may be worth carrying a spare set with you on your detecting trips.

OTHER ACCESSORIES TO CONSIDER WHILE METAL DETECTING

A finds' pouch/coin & artefact box and a water/drinks bottle – in the event of finding anything of interest then it is crucial that you have something with you in which to store it. A small plastic box with inserts will be enough to hold any coins and artefacts you find and adding sponge padding inside will offer the protection these finds deserve. They can be easily obtained: anglers with their fly and tackle boxes use something very similar. Finds pouches are also easily purchased and can be wide ranging; workman's tool belts have been used, also the metal detecting manufacturers sometimes make their own and add them into the deals they offer on their machines. Worn around the waist they come with different compartments to store the variety of objects that you may find, from the undesirable rubbish, such as ring pulls and empty drinks' cans, iron horse shoes and so on, or some of the fantastic coinage and artefacts that exist in the soils. All will need to be kept somewhere and the finds pouch is the ideal accessory. Other compartments can be used for holding a drinks bottle to quench your thirst while searching on hot days, there is nothing worse than being out in the scorching heat only to realise you have nothing to keep you hydrated, so take a small bottle of water or juice – enough to keep you on the fields longer.

A finds pouch attached around the waist.

CHAPTER TWO

Technique: How to Use the Detector Properly

WAY BACK IN the 1970s, I had my first adventures with a metal detector, but being a child I didn't really understand how to use it correctly, and with limited concentration levels it's no surprise that I found very little. Fast forward a little to the 1980s, when I bought what I classed as my first serious detector; using this machine every week for the best part of five years until I replaced it, meant I developed a successful technique that was to stay with me until the present time. My three basic rules around technique are simple and straightforward, and I would like to share them with you.

Firstly, make sure you have the detector's search coil as close to the ground as possible. You need to imagine you are feeling every blade of grass with it; the lower to the ground you scan the more your chances of finding something of interest increases. Over the years I have detected on farmland here in the UK with a number of different individuals, each with their own technique. The ones who scoured the ground in the manner I suggest were usually the most successful at finding things, But to my utter dismay I have also seen people detecting with their search coils so far off the ground that they have little or no chance of locating anything of note. This in my view is a complete waste of time, you may as well switch off the detector and just enjoy the fresh air!

Next up in the quest for having a successful technique is speed, or lack of it. It makes perfect sense to me to detect as slowly and as methodically as possible; once again I have seen people literally running across fields swinging their detectors with very little chance of locating anything, while at the same time others walking behind them across the same section of land in a slow and methodical manner pulling up a whole host of different coins and artefacts. The rule is simple, the faster you go, the less you will find and the more you will miss: low and slow is the key! The sweep speed must be slow, keep the coil flat to the surface and at the end of each swing don't lift the coil, keep it flat, even at the end of each swing. That extra few inches of ground coverage could be the key to coming home with something desirable.

Now that you have your technique sorted it is important that you overlap each sweep. Concentration is needed here, as after a few hours the overlapping could become a little shabby, but by overlapping each sweep you are simply covering all the necessary ground in front of you and maximising your chances of making finds. Keep to these simple yet effective rules then you won't go far wrong; it's too easy to fall into the trap of speeding up and going fast to cover more ground but please resist, you will find much more not doing it.

A tip here is to periodically check the search coil isn't loose; sometimes if the coil bolt isn't tight it can cause the coil to move around and be positioned at certain angles. Educate yourself to check the coil for looseness every hour or so as this simply helps keep that coil in the correct position for your technique to work. However, it is worth noting that you should not over tighten the coil bolt; doing so can apply pressure to the lugs either side which could lead to them both breaking. These little things can make a big difference out in the field.

CHAPTER THREE

Gaining Permission to use your Metal Detector

SO YOU HAVE your detector! You should also have all the relevant gear to start the adventure, but where do you begin?

Firstly, you have to decide where you would like to search. Will it be the beaches, or will it be on land, such as farmers' fields, woodlands or footpaths? Once the decision is made the next step will be seeking permission to search such areas.

SEARCHING ON LAND

Please remember and make a mental note if you can that someone, somewhere, will own the fields or land that you would like to search. No matter if it's in the middle of nowhere, if it's an area of land that you have walked or driven past every day and never once seen anyone remotely linked to it, worry not, it will be owned by a farmer/landowner, a Trust or a local authority, and it is your duty as a responsible detectorist to find out who and approach them to see if they will allow you access to carry out your search. This procedure can take some time; not all land is owned by local famers; I have had permission to search arable land in the past where the landowner actually lives in a different county. The first port of call will be to visit the nearest visible farm in that area; if you are lucky they will be the landowners, if not, they should be able to point you in the right direction and give you the information that you require to secure the permission that you need. My advice here could be crucial in you securing your first ever area to search and I would advise against obtaining a phone number for the landowner, unless they are based many miles away. The rule of thumb should always be to meet them face to face, although meeting someone for the first time and asking them if you can use a detector to search their farm can be quite daunting. Please remember they can only really say one of two things, yes or no, but if you are polite and come across in a respectful manner there is every chance they will say yes. It might also be worth taking your detector with you, so they can see the device that you are using; this may help them understand the hobby a bit more and what you are trying to achieve. Be prepared for some to respectfully decline your request, as will undoubtedly happen and don't let this put you off asking others in the future, it only takes that one kind landowner to grant permission and then who knows what adventures lie ahead.

There will be many reasons why a landowner would decline your request: it could be simply that there are already other detectorists searching the land, and their relationship is solid; it could also be that they don't like the idea of someone digging holes in their fields and this is an area that you can discuss face to face with the farmer when asking. Most objects are buried shallow so it will be worth

A stubble field ready to be detected!

mentioning that you won't be necessarily digging deep holes. Some detectorists use stainless steel gardening trowels, especially on arable land – mentioning that you will be using one might well dispel any fears the farmer might have with regard to what you would be using to dig on the land.

At this point I would like to impart some really useful information to the reader. It is based around a method I used when I first started serious metal detecting and it has helped me ever since. Start small: horse paddocks, smallholdings, even fields or areas of just a few acres will do. This will not only help you learn how to use your detector and get used to its features without feeling pressurised by wandering over vast acres wondering where to go next, but small parcels of land will be easier to obtain permission for than say an 800-acre estate, that other detectorists have visited over the years, with similar ideas to yours. Firstly, think about approaching friends and family; ask them if they know anyone with a paddock or a smallholding, someone with a small patch of land who would be willing to let you search. Ask people at work and don't be afraid to do so, the worst they can say is no. There are many other ways to potentially gain permission; sharing just some of the methods here should put you on the right track. Eventually someone will pass over details to you that will lead to you gaining your very first permission. Once it is obtained you can later use it as a reference and an example to show other landowners in order to gain more permissions,

However, human nature being what it is, you may simply want to bypass the smaller areas and target the larger farms and estates. There is nothing wrong with that, but please remember that landowners and farmers are extremely busy people, usually up at the crack of dawn tending to their livestock and other agricultural duties. It may be wise to visit just after lunchtime when he or she probably has more time to speak with you directly; also worth noting here is that the larger farms will also have been noticed by other detectorists, especially those with historical landmarks nearby. This may work for or against you depending on the landowner's outlook on things; he may be happy for you to share the permission and fields with others or point blank turn you down as a lone detectorist or a small group is enough for them to search their acreage.

Again based on personal experience and my opinion, never plead with a landowner/farmer. If no is their answer then take no as their answer! If you continue to put your case forward, or, heaven forbid, argue against the decision, it could well put them off granting permission to a detectorist including yourself in the future; so be polite, take it on the chin, you never know, a return visit in twelve months' time may result in a different outcome.

One particular example of this stands out in my mind. A number of years ago I was turned down twice in successive years for permission on a farm which had history dating back a thousand years or more. The landowner had no interest in history and didn't warm to the idea of having anyone on his fields searching for lost coins and artefacts. Just by chance one morning I was detecting on his neighbour's land and the weather had turned wet. I was taking a ten-minute break from the rain under a tree along the roadside when a motorist had stopped to take a call on his mobile phone. A few minutes later the vehicle tried setting off only to find its rear wheels had become stuck in the roadside verge and within seconds both back tyres were half buried, I offered my assistance and eventually helped push the car out. The driver turned out to be the landowner whose farm I had been trying to gain permission to search and a brief conversation resulted in him granting the whole farm for me to search. I'm still searching the farm almost a decade later.

At this stage, presuming permission is granted, it's vital that I mention one important factor: please leave the fields that you are searching as you found them. There is nothing worse than hearing stories of detectorists too eager to get onto the land, leaving gates wide open, climbing over fences, potentially damaging them, and of course not filling their holes in correctly. If the landowner decides to walk out into the area that you have been searching and notices holes partially filled in and not stamped down correctly then chances are that permission will be withdrawn. It is common sense to leave everything as you found it. Treat the land as if it were your own and you won't run into any problems or issues.

While on the subject of landowners, it is always advisable to show them what you have found after a day's detecting; you may not think it's worth showing a handful of lead, buttons and copper coins for example, but some are extremely interested in the history around their home. It is likely that they will be happy to take some of the coins off you to show the rest of the family. It also shows that you are trustworthy and this could potentially lead to permission for you to search more land in the future. Sharing is the way forward!

SEARCHING ON BEACHES

This next section looks at permission for metal detecting on the beaches; just as popular and rewarding as searching in fields for those willing to spend the time searching this environment.

You can search either at low water or when the tide is full, and with over 7,000 miles of coastline in the UK – and a lot of it available to search – you can see why it holds an attractiveness all of its own. Gaining permission is fairly straightforward; for a number of years a Crown Estate foreshore licence was available for people to download from the Crown Estate website, which was completely free of charge. However, at the time of writing the licence is currently unavailable. That said, the law remains the same: as long as the beach is owned by the Crown then it is fine to detect.

The following is an edited extract from the Crown Estate website:

> *Anybody wishing to carry out metal detecting on our foreshore (defined as the land between mean high water and mean low water) may do so without a formal consent from the Crown Estate. This permissive right applies to Crown Estate foreshore only and not to the seabed, river beds or any other Crown Estate lands. Please be aware that there are a number of private beaches throughout the UK which are not accessible and permission must be obtained from the owners if people are thinking of searching in these areas. As with everything in the hobby of metal detecting, common sense plays a major role and the beaches must be left in the same state as when you found them, holes all over the sand for example will draw attention, and even though the tide may well flatten them out it is always best practice to make sure they are covered correctly and neatly.*

Beaches like this will always hold discoveries.

CHAPTER FOUR

Where to Look: Taking Time to Find the Right Areas to Detect

I HAVE BEEN particularly looking forward to this chapter because spending time researching places and areas to use your detector with a good chance of unearthing potentially old coins and artefacts is just as rewarding as actually finding them; for me personally it is the one of the most rewarding parts of the hobby. The sheer number of reference books I have purchased from new and second-hand book dealers could easily have filled an average-sized library, looking for snippets of information that could potentially lead me to a field, a footpath or an area that in the past that had considerable foot traffic, where many people walked, worked or spent leisure time. Inevitably over the years, losses made their way into the ground, so keep the words 'foot traffic' burning brightly when it comes to research.

SEARCHING FOOTPATHS, FAIRS AND FIELDS

Take old villages for example – many of them had an area or field which once a year held events: these could be for fairs or country sports and would have been a highlight in the lives of hard working people in the past; a place where they could buy, sell, barter, gamble, meet old and new friends from outlying villages or towns and generally enjoy themselves. Some fairs ran for more than one day, 2-3 days is not uncommon and can date back in time a few hundred years at the very least. Now imagine if you could locate the actual field(s) where these events were held; imagine the potential coins and artefacts you might discover. Well many people do just that: locate the actual area, gain permission and uncover some really choice finds.

Reference books from the library written by a local historian often mention pastimes or sporting events in bygone days and such enlightening books as these are pretty accurate in placing the location of such events and, providing they haven't been built on by modern development, could lead you to your first really good finds. Locate the landowner and once permission is granted (fingers crossed) don't be shy in asking if they know the location or history surrounding them, chances are they will and it will give you a good indication as to what you are likely to find age wise. Some of these events and activities dated back to the medieval period and continued right through to the Victorian era when sports and pastimes became very popular.

I touched on footpaths earlier as being very interesting areas to detect. Most paths date back hundreds of years and the foot traffic over that period will have been considerable; for example a path leading from a village to the local church will have seen people making that journey several times a week for countless decades, so you can only imagine how many coins and artefacts will have been lost as they walked. Even now if I am lucky enough to pick up a new permission, any footpaths crossing the land and fields will be searched first. Purchase a detailed Ordnance Survey (OS) map of the area that you are searching – the map will highlight the footpaths that crisscross the fields and give you an idea where they begin and end – but don't worry if they seem to be out in the middle of nowhere. They will have all been used considerably in the past, taking people back and forth to work, short cuts, moving livestock and a whole host of other activities. A handy tip here too: if you find

Old footpaths are always worth a search.

the modern, detailed maps a little expensive you can find an abundance of maps covering most or all counties in second-hand shops and charity outlets. Some may be a few years old, but since the footpaths have been around longer than the maps, they will be marked on even second-hand or older maps. And you'll be saving your hard-earned money at the same time!

Staying on the subject of footpaths, it is also worth searching either side of them, since people didn't necessarily walk in single file; the route may well have seen crowds walking three or four abreast in some cases over the centuries, so it pays off to search the wider surrounds. I recall a few years ago finding a lot of coins on slightly raised ground just off the main track, which in the winter becomes flooded in parts. My guess is people walked on the harder, drier terrain to avoid getting their feet wet and this is one example of thinking about how our forebears may have used their common sense. Use that insight while out searching and looking for clues to where discoveries may lie.

Quite often carrying out simple site research will open up more questions than answers. The following example will illustrate this. A particular farm I visit occasionally clearly shows a footpath that winds its way through four of the landowner's fields; nothing unusual there except on closer inspection there are actually three other potential pathways that meet in one random place. This was only highlighted when I scrutinised early maps while searching for other features, so I would recommend that once you have the area you would like to search mapped out it can pay dividends to look at much earlier maps, such as Victorian examples. Some of these are highly detailed and show many features such as wells, windmills, long gone churches, cottages and farms.

Having discovered these new footpaths, I returned to the particular field at the first opportunity and found many choice coins and some artefacts, many dating from medieval times. Why these pathways met up at one point I am unsure; perhaps it was an important meeting place, or an ancient religious place – now lost in time. Either way, it is an example of detecting posing more questions than answers; something you will find happens a lot.

One of the best sources of early maps is National Library of Scotland's (NLS) 'explore geo-referenced maps' which details all areas of the UK and is easily found online at https://maps.nls.uk. I have been using it with some success over the past five years and it has led me to many new discoveries. If you use a computer and are thinking of using its online capabilities to help search out potential new places and maps then the NLS site is well worth saving as a bookmark. While we are on the subject of using modern technology for research purposes, it is worth downloading Google Earth to your laptop or PC. Based on satellite imagery, this computer program will open up a whole new world of features, footpaths and sites for you to explore. Spending a little time looking down on the places you want to detect could pay dividends and it is a tool that I now use on a regular basis: it's amazing how many times just by looking down at a group of fields 'from the sky' I find I have previously missed a feature or footpath – even in fields where I have found coins and artefacts only hours before. Take a little time to learn its handy features and study the information on the toolbar; simply zooming into what at first seems to be a featureless field may reveal far more detailed information than wandering around aimlessly on the ground.

OTHER AREAS OF RESEARCH AND POTENTIAL PLACES TO SEARCH

When I am out detecting, especially a new field or farm, I always look for physical features: depressions in the ground, unusual humps and bumps, overgrown ponds and old wall systems to name but a few. This can lead to unearthing some really interesting finds.

Take the overgrown pond in a field for example: drovers, farmers and shepherds will have used the same pond for watering their cattle and livestock; some ponds go back many centuries and were used extensively. The banks and circumference of the pond should be searched methodically as discoveries will undoubtedly exist.

Depressions in the ground are an interesting feature to look at. Many years ago a friend and I were lucky enough to gain permission from a landowner in Cumbria, near to where an ancient Peel tower had stood for many centuries. The nearest pasture field to the tower had an almost perfect circular depression imprinted on the land, which at first glance we thought may well have been a drained pond. However, as we walked over to it we realised it was just too small to be a pond. As we started searching around it, I headed directly to its centre, but a scan with the detector produced very little. However, my friend searched around the circumference and immediately unearthed a lovely hammered Elizabethan silver sixpence dated 1575. The coin was so good we stood and stared in awe of its condition for some time, wondering who owned it and how it had been lost. A little time later, my friend found another similar coin of the same period, and went on to find another three coins, dating from the reigns of Charles I and James I, all at the edges of the depression. We then realised that this might be the site of a Tudor cockfighting pit and this was confirmed some time later when I found reference to the very place in an old book written at the turn of the last century. Under a section named 'cockfighting in bygone days', it told of a pit being in existence in that particular area, and more of these pits were mentioned throughout the county, which we tried in vain to track down. We did manage to gain permission to search another pit, but a local detectorist had beaten us to it some time earlier. We revisited the scene of this rather gruesome pastime on many occasions in later years and did manage to winkle out more coins, but eventually the fields became part of a larger estate and the whole area became out of bounds for the detectorist. I still think back to our adventures there, and wonder what we might find using the modern, more technically advanced detectors that we see and use today.

SEARCHING RIVERS AND STREAMS

Rivers and streams have drawn people to them for countless years: fishing, poaching, swimming, relaxing, or just generally messing about on the river. You can almost guarantee that you will make coins and artefact discoveries here, due to the many activities that

Riverside locations will be worth detecting.

have taken place. Once again it pays to scrutinise old maps, as many rivers in the UK had fords which could be crossed by wading the shallow water. Some fords are natural and became ancient routes and some were constructed by man; many will be clearly marked in bold black on the old maps and you will be surprised just how many exist. The drovers will have regularly used these watery routes, stopping off to let their cattle and livestock rest and drink, and in addition, the routes and tracks leading to and from such fords will be well worth searching with your detector; with the amount of foot traffic it is not surprising that many desirable finds have been made over the years in such areas.

SEARCHING ROADS AND HIGHWAYS

How can we discuss potential places to detect and discover without mentioning roads? The word 'road' goes way back, probably a lot earlier than our history books state; however, we can get a pretty detailed grasp of the name from old English: 'Rad' riding' and Middle English: 'Rode, Rade'. There is reference to Rad linked to 'Riding on horseback' so you can see and understand how the word and meaning fall into place. A road linking two places together was different to a footpath; carts, horses and other vehicles will have carried goods and produce many miles using a road. They also carried armies and their supplies, by cart and foot, marching many miles to reach their destination; you only have to look at the Romans and their advanced road networks branching out like a spider's web across the UK and Europe to see roads were a vitally important part of everyday life back then. But how do we search roads? Most of them now have a tarmac surface and are very busy, potentially dangerous places to be, but not all – some Roman roads can be visible as far as the eye can see, and gaining permission to search farmland either side of the said route can considerably increase your chances of finding your first Roman coins or artefacts. Detecting on the Roman roads themselves is not advisable but scanning the land as close as you can get will lead to decent finds. Some of my most desirable Roman coinage has been unearthed in the vicinity of these ancient routes, tracing these networks and doing your research into where they lead to and from can be some of the most exciting and rewarding aspects of the hobby, but please ensure permission is granted by the landowner first before venturing out to detect near to them.

But what about roads other than the Roman ones? Well the

Ancient roads between villages hold finds!

modern day routes will have, in many places, replaced existing ones, some dating back many centuries. Roads leading from villages to towns were always busy routes and searching in the fields either side of these can prove to be very rewarding. One example of my own search permissions has fields that border a route going back a thousand years or more. Although now it is a modern day, extremely busy road it has remained pretty much in the same line and direction as it was in the past and getting as close to the road and hedgerow as I can often produces medieval pennies, medieval artefacts and discoveries from later periods. Of course, searching this close to a main road has its issues: noise from the cars and lorries is one problem and the sheer amount of scrap waste that has been thrown out of car windows can be another, but if you persevere the finds will come to light. Once you clear out the modern scrap, bits of tin cans, ring pulls and foil, you should have a pretty clear run at the older objects hidden there. Another tip here, always wear headphones in this environment, it will blank out any traffic noise and enable you to listen properly for the signals that count.

STUDYING ENGLISH PLACE NAMES BEFORE SEARCHING

One method of research I used for a long time with moderate success was studying English place names. This is a really interesting part of the hobby and I will explain why. When you are travelling around the country, to various towns and villages and you see road signs with place names on them, have you ever wondered what their meaning is? How did they get that name and were does it originate from? Well I certainly did and the findings were very interesting.

For example a permission was granted by a landowner near to Broughton in Cumbria, which turned out to be a common enough place name throughout this country, but where did it originate, how old was it and would it give me any indication on what I was likely to find there? In a trip to the local library, I soon picked up a reference book for the village, which detailed the age of the church, mentioned certain houses and revealed that the name itself, 'Broughton', was a common name usually meaning 'farmstead by a brook'. This particular village was first mentioned in 1196 (according to the book), which indicated to me that the medieval folk were living, and well settled in that village. The river area turned out to be the first area worthy of a search and over time produced finds from that period. So in a nutshell, place names can give you some really interesting clues to certain areas that you may wish to detect so find out what the village name means and when it dates back to. It is worth remembering also that in England alone there were 13,418 towns and villages mentioned in the Domesday Book, written in 1086. This gives you some idea of how vibrant and rich in history this country really was back then.

SEARCHING BEACHES

In many cases, beaches are the first place people search with their metal detector; as I mentioned in the previous chapter there are over 7,000 miles of coastline in the UK, most of which is accessible to people wanting to give metal detecting a try. Again we think about foot traffic, and beaches, especially in the summer months are a haven for people enjoying themselves, sunbathing, relaxing, swimming, fishing, the list goes on; so it isn't hard to see why beaches can and do hold a lot of potential finds for the detectorist. There are two different environments to search on beaches: the wet sand, low tide area and the sand and shingle section from the high water mark to the start of the wet sand; both are equally rewarding, but you need to know which hotspot areas will reap the potential rewards. Looking back, when we ventured off to the beaches with our metal detectors, my friends and I pretty much searched everywhere and anywhere at the start, but with experience we soon picked up where the best areas were to look; so for example when the area from high water down showed signs of bare shingle patches we knew coins and jewellery would come to light, mainly modern coins that had probably been lost by seaside tourists and sunbathers, with the odd piece of jewellery mixed in. It kept us busy, sometimes coming home with as many as fifty coins each, some spendable, some older. However, with beach detecting, the glory can be short-lived as the

A typical detecting beach scene.

very next day, depending on the tides and the wind, these shingle patches can simply disappear, buried under a carpet of rocks and small boulders. It was all about reading the beaches, and we soon became moderately successful at doing it. The direction of the wind played a big part, south-westerlies for example seemed to strip the beaches of the rocks to reveal these shingle patches and northerly storms did the same, stripping the carpet of boulders overnight and we would be ready to take advantage of these conditions the very next day. Soon we found that a pattern of easterly winds arriving didn't make any impact into the beaches we searched, sometimes these weather patterns would stick for weeks on end and left us feeling a degree of frustration. When this happened we would head to the wet sand, where, more often than not a gold ring or two would appear; in fact wet sand detecting was more productive, certainly for gold and silver jewellery.

If you think about it for a moment, it is easy to see why so many rings are lost at the beach, especially on the wet sand; swimming in cold water can quickly numb the hands, leading to cold fingers where the rings are liable to slip free, sometimes with ease. It is not unusual to see seasoned beach detectorists finding dozens of gold and silver rings over a season, this is on top of many coins and other pieces of jewellery too. However, wet beach detecting differs from searching on the dry sand and shingle and commands a different type of metal detector to cope with the high levels of salt and mineralisation that occurs in that environment. The same single frequency detector that you may have been using around the high water mark may not cope on the wet stuff, in fact the instability is likely to drive you crazy! The science behind it is fairly easy to understand: when the salt becomes soaked it becomes highly mineralised; and sometimes this can be also mixed with something called black sand, which is also blended with ingredients like iron for example. Put a highly sensitive metal detector over it and the machine will be sounding off all over the place, not an enjoyable experience. In some cases it is nearly impossible to use without lowering the detector's sensitivity; by doing that you may make the detector slightly more stable, but you will of course lose depth and this then defeats the object of searching the sands in the first place. Off you trot home feeling very frustrated.

The remedy is simple and straightforward enough: if you are serious about wet sand searching then concentrate on two types of metal detectors, the first being pulse induction machines and the other multi-frequency units because both choices can cope with the minerals and allow you to detect with relative ease. The added advantage of the multi-frequency platforms is that you can also use them with great effect inland, making them a true all-terrain detector; the pulse induction machines are built purely to cope with the beaches and once mastered are extremely efficient at finding things, partially due to the extreme depths they can reach. There are many different types of these units so spend a little time researching the costs and features before you decide which to buy. I would advise you to look carefully at the weatherproof, waterproof metal detectors since you are searching in a wet environment and it makes perfect sense that the unit is protected, not only from

A likely looking beach to detect.

dampness but from the potentially damaging effect salt has on components. In my opinion there is nothing worse than spending your money on something that might corrode and let you down after a short period of time.

THE BEST BEACHES TO SEARCH

Which beaches that are most likely to produce finds? Remember not all beaches were popular, some are very remote with very little foot traffic over the centuries and probably won't be worth spending any serious time scanning, but with common sense and a little research you can easily find the places where, for example, the Victorians, spent their recreational time. The Victorians often only had one day off a week and loved nothing more than to spend it at the beach. With this in mind, one area of research that I found really useful was old postcards; these are important glimpses into the past and often show large crowds of holidaymakers and locals enjoying the sun, sea and fresh air. Some postcards unexpectedly revealed popular, packed beaches, ones that are relatively quiet nowadays but were very busy a hundred or so years back. The images also revealed parts of the beach where people flocked in larger numbers, maybe there was a nearby drinking fountain, a bandstand or simply a vantage point to hop on and off the trams that were used to transport people back and forth. With this in mind, we usually concentrated on the beach area below these landmarks and, more often than not, it paid off.

These vintage postcards are easy to come across as they can be found in large numbers at coin and collectable fairs, antique stalls in village halls and other such outlets. Not only do such cards reveal hotspots for beach detecting they also show images of inland sites too, such as Victorian maypole events, shows, fairs and agricultural gatherings. All in all they are a really important part of site researching, and one I've taken advantage of over the years.

Portrush County Antrim, in the early 1900s.

It is also worth pointing out that there are many online links and images showing postcards for sale and you can study those showcased by sellers. Also check out certain Facebook and other social media pages dedicated to bygone days that show many pictures

Wet-sand beach finds.

of postcards and photographs from over the last century. Join a group showing the area you are interested in searching and you will find over time that particular social media platform will be of great interest to you and your quest for researching potential detecting sites.

Now we can take a look at other specific areas of the beach to search. We have already discussed the dry sand and shingle section so now let's look at the low water/wet sand search environment. Reading the sands is pretty much crucial to your success, learning what to look for and what to avoid is vital. It is also worth remembering the sands can move on a daily basis: one day you may have a rock patch or gulley to search, the next day this will have disappeared completely. Keep your eyes peeled for both of these areas as rings and coins will be in detectable range, held up by the stones. Rock pools can be lucrative and as soon as you notice them give them a scan, the sand, however, will quickly cover and storms will uncover them – it is a continuous cycle.

Let's not leave this section without mentioning safety: you must be aware of the tide times and when the tide is due to turn. There is an old saying that I used to have constantly going around in my head: 'Tide and time wait for no man', so buy a tide-time booklet or search online for the tide calendar. Incoming, fast tides can be very dangerous and can easily catch out people who are concentrating on other things; too many lives have been lost because of the sea, make sure you stay safe!

So to recap: rock pools, gullies, stone scars and pebbled areas are all places you should be searching as soon as the tide goes out. Reading the beach and tides will come second nature to you once you start finding.

CHAPTER FIVE

The Law and Treasure: What you Need to Know

MANY HOBBIES HAVE rules, regulations and laws, and metal detecting is no exception. This chapter will cover all the information you will need to know as defined by the Treasure Act 1996. (Please note this Act does not cover Scotland or the Isle of Man.) The chapter also includes details of the Metal Detecting Code of Conduct and the Portable Antiquities Scheme for England and Wales.

While enjoying the hobby of metal detecting, at some point you may well stumble across an item of treasure; that could be in the form of a gold or silver ring, a piece of old jewellery, a pendant, brooch or dress decoration, a hoard of coins, either scattered or inside a vessel or pot, the list goes on. It is important to know what steps you need to take should this happen. Firstly, you need to be sure the find is actually a treasure piece; if you are uncertain you can seek out the local Finds Liaison Officer (FLO) linked to your county, a simple search on the Internet will reveal who the FLO is and their contact details. Some may take some time to get in touch, and if you feel you have not had a full reply you may try emailing the British Museum (treasure@britishmuseum.org) with the query; they will ask for details and some images of the find to rectify whether or not it qualifies as treasure. Some individuals may want to post their finds on social media platforms to obtain information on their item; this can be a good idea as long as the information coming back is correct and many seasoned metal detectorists do use social media such as Facebook 'detecting groups' as a way of helping them identify their finds.

A medieval hoard of coins found whilst detecting.

TREASURE LAW

At the time of writing, the law states that any metallic object with 10 per cent or more gold or silver content will be classed as treasure, providing it is over 300 years old at point of discovery. (This excludes single gold & silver coins.) If the artefact is prehistoric (Iron Age or earlier) and any part of it is either gold or silver, no matter what the percentage is, it will be deemed as treasure. If two items of prehistoric date are found together then these will also be deemed as treasure, regardless of what they are made from (copper, bronze etc.). An example here would be an Iron Age sword in its scabbard, which counts as two items and would therefore be classed as treasure. This is all fairly straightforward; however, under the Treasure Act there are further matters to consider.

COINS

The law states that a group of coins (two or more) of precious metal, from the same discovery and at least 300 years old will fall under the Act. However, it is important to know that if the coins contain less than 10 per cent precious metals then the number of coins needing to be declared rises to ten; an example being if you are lucky enough to unearth a concentration of ten or more Roman bronze coins then the find must be declared; if the number is fewer than ten then you and the landowner can record the finds and, depending on the agreement you have, may decide to keep them for your own collections. Other parts of the Treasure Act include hoards of coins that have been purposely buried and hidden, small groups of coins that have been dropped or misplaced (this would include purse losses and spills) and both ritual and votive deposits. Interestingly, any items that are less than 300 years old with a high percentage of precious metals in their make up, and which have been hidden away with the intent of recovery will also fall into this category.

DECLARING POTENTIAL TREASURE ITEMS

Should you be fortunate enough to find potential treasure as listed above, where would you report it? It is fairly straightforward and not as daunting as some might imagine.

First you must report the discovery to the District Coroner or the local Finds Liaison Officer within fourteen days of finding

The gold Saxon pendant made from a Roman coin!

it. Personally I found the process of contacting the coroner was easier and quicker. Once this action is complete you can relax, you have carried out what is required by law, the local FLO should then contact you to talk about where and how you made the find and to organise a meeting which mutually benefits both parties and to pass the find over. You will also get a receipt of the find once it has exchanged hands.

At some point the officer will write up a report on the item, and you will find this information of great interest, as you will get an accurate date and description of the object, along with other relevant details. After this procedure there could well be a period of waiting, investigations into some treasure items can take over a year to come to a conclusion. My advice here is just be patient, let it go through its official action, and view it as something to look forward to.

I did exactly that when, in February 2013, I managed to find a wonderful Roman gold coin with a gold loop attached at the top. This loop turned it from being a coin into an artefact (pendant) and that meant it was now classed as treasure. I drove back to the farm to show the landowner, quickly discussed the Treasure Act and the process behind it and that was that. The week ahead I arranged to meet the FLO, who was actually rather underwhelmed when I handed it over and I got the impression she imagined it was going to be the size of a dinner plate rather than something the size of a modern pound coin! Either way this part of the process was now complete and I had to wait until information came through declaring it treasure by the District Coroner. The whole process took around eighteen months from start to finish and I did wonder at times if it was ever going to be finalised, but now I realise that the people involved have busy schedules and the treasure process isn't just about my one find. Eventually it will come to a conclusion – whether or not you agree with the outcome. In this case, a museum wished to acquire the find, the reward was agreed and split between myself and the landowner and a few months later a cheque for the amount landed on each of our doorsteps and that was that.

As mentioned earlier, the Treasure Act 1996 does not apply in Scotland or the Isle of Man.

All archaeological objects found north of the border should be reported under the Scottish treasure trove law; for more information on the law in Scotland you can search online for **www.treasuretrovescotland.co.uk** or alternatively you can write to:

Treasure Trove Unit
National Museums of Scotland
Chambers Street
Edinburgh
EH1 1JF
Tel: 0131 247 4355
Email: info@treasuretrovescotland.co.uk

The laws in the Isle of Man are also different, and it is important that we mention it: all archaeological objects found on the island should be reported to the National Heritage Agency. For more information you can contact:

Manx National Heritage
Manx Museum
Kingswood Grove
Douglas
Isle of Man
Tel: 01624 648000
Email: enquires@mnh.gov.im

For detectorists in Northern Ireland there are specific details

Treasure discoveries from Cumbria.

regarding the Treasure Act and also the antiquities law, something which you will need to become familiar with. Many great discoveries have come from this part of the country so it is vital that people understand the laws. For more information contact:

Department for Communities
Causeway Exchange
1-7 Bradford Street
Belfast
County Antrim
BT2 7EG
Tel: 028 9082 9000
Web address:
www.communities-ni.gov.uk/articles/finders/treasure

CHAPTER SIX

The Metal Detecting Code of Conduct

IN THE LAST chapter we concentrated heavily on the laws and procedures surrounding metal detecting; in this we have a detailed look at the metal detecting code of conduct. So, what is it? And how was it brought about?

The hobby has a representative body of elected volunteers, the National Council for Metal Detecting (NCMD), who, some time ago put together a code of conduct for all metal detectorists to adhere to. It is simple and straightforward and offers the advice to individuals who may be a little unsure about certain aspects of the hobby.

The advice in simple terms is as follows:

1. Do not trespass. Obtain permission before you venture onto any land.

2. Respect the country code, leave gates and property as you find them and do not damage crops, frighten animals or disturb nesting birds.

3. Wherever the site, do not leave a mess or an unsafe surface for those who may follow. It is perfectly simple to extract a coin or other small object buried a few inches below the ground without digging a great hole. Use a suitable digging implement to cut a neat flap (do not remove the plug of earth entirely from the ground) extract the object, reinstate the grass, sand or soil carefully, so that even you will have difficulty in locating the find spot again.

4. If you discover any live ammunition or any lethal object such as an unexploded bomb or mine, do not disturb it. Mark the find spot carefully and report it to the local police and landowner.

5. Help keep Britain tidy, dispose of any refuse you come across.

6. Report all unusual historical finds to the landowner, and acquaint yourself with current NCMD policy relating to the voluntary reporting of portable antiquities in England and Wales and the mandatory reporting in Scotland.

7. Remember it is illegal for anyone to use a metal detector on a designated area (e.g. scheduled monuments (SM), Sites of Special Scientific Interest (SSSI), or Ministry of Defence property) without permission from the appropriate authority. It is also a condition of most agri-environment agreements that metal detecting access is subject to certain rules and regulations including mandatory finds and reporting of all finds to the Portable Antiquities Scheme. (As stated on the NCMD code of conduct web page.)

8. Acquaint yourself with the terms and definitions used in the following documents: Treasure contained in the Treasure Act 1996 and its associated code of practice, making sure you understand your responsibilities.

9. Remember when you are out with your metal detector you are an ambassador for the hobby. Do nothing that might give it a bad name.

10. Never miss an opportunity to explain your hobby to anyone that may ask about it.

For a more detailed look at the laws surrounding metal detecting it is worth spending a little time looking online for the portable antiquities website, which contains a lot of easy-to-understand information. There are also pamphlets available in some museums that can be taken away and read at leisure.

THE NATIONAL COUNCIL FOR METAL DETECTING (NCMD)

The NCMD was created in 1981 to provide a forum for responsible metal detectorists to discuss issues that may affect the hobby, and to be an authoritative voice to counter ill-informed and frequently misleading criticism of the hobby. It is worth noting that the NCMD does not represent the trade or archaeological interests.

The NCMD also gained government recognition as an organisation which represents metal detector users nationwide; importantly it has played a major role in representing the views of the detectorists to government departments regarding legislation affecting the hobby. In terms of any future legislation, they will undoubtedly represent the views of metal detectorists against any changes to laws that could potentially damage this wonderful pastime.

Further information is available on the NCMD website, ncmd.co.uk, including details regarding the insurance it provides. Public liability insurance (PLI) is a handy thing to have, especially if you are thinking of attending metal detecting rallies as most, if not all, of these events make it compulsory for the attendees to have this in place. PLI will cover you for most mishaps you might encounter while out in the field. Other useful information provided on the NCMD website includes: search agreements between you and the landowner/farmer, which you can download, details of regional representatives, and much more.

NCMD
NATIONAL COUNCIL FOR METAL DETECTING
TO PROMOTE
TO PROTECT
TO ENCOURAGE

CHAPTER SEVEN

How to Record your Finds

OVER THE YEARS it is hoped that you will uncover many interesting coins and artefacts with your metal detector, after all unearthing and saving items from potential destruction and bringing them back to life, so to speak, is a big part of the hobby and is the main reason why we take part. I have lost count of the many interesting and sometimes unique items I've seen and held that detectorists have brought to light; some I've even found myself – a few of these being classed as treasure.

A medieval key unearthed by one lucky detectorist!

All are potentially of historical interest, so with this in mind it is certainly worth considering recording the items you find. Some items may seem to be broken pieces of scrap, but many have a story to tell and you can gain so much information from that one find you have made.

As mentioned earlier, the port of call for recording your metal detecting finds is the Portable Antiquities Scheme, which was started in 1997 as a voluntary programme by the UK Government and covers England and Wales. Its purpose is to record the increasing number of finds made by private detectorists that would otherwise go unrecorded.

The scheme itself funds the posts of the Finds Liaison Officers who are mainly based at local museums or at county councils, where the detectorist can arrange a visit to report their discoveries. The officer is qualified to examine each find and provide you with more information about it, they will then record the find for you and (most of the time) give details about the item such as its function, the metal it is made of and the date of the period it came from. Sometimes the information may be incorrect as it can be a difficult task to accurately determine the age of certain coins and artefacts, especially those that are broken or worn.

As this is a voluntary scheme there is no pressure on you to get any of your items to the officer, and I have found taking detailed images and sending them to the FLO by email a really efficient way of getting the initial information to them. You will need to include details of where the object was discovered – a town, village, parish etc. – and some objects may need to be handed over in person so that the officer can make a better, more detailed examination. You may also be asked if the object can be kept for a period of time for further analysis – for which you will be issued a receipt, while it is in their care. (If one isn't issued, make sure that you ask for one.)

At this stage you may still be wondering what you will gain from recording your finds. The FLO should offer you a wide range of services, not only detailed identification but also advice on conservation and storage;, advice on the Treasure Act and information on the historical importance of the find. The officer will also provide you with a copy of the information they have recorded. Once the items have been recorded, they will be published on the portable antiquities online database: www.finds.org.uk – you can register free of charge to this scheme at any time.

Over the years I have recorded many finds, mainly ones I wanted

Silver hammered coins can be plentiful in certain counties.

to learn a little more about, or finds that I struggled to identify myself. I found the process easy and helpful, however, I did list a group of finds that took a few years to be recorded. This did not worry me and I forgot about them completely until a batch of emails turned up with details of the items attached, so it might be worth asking how long the process will take, especially when you hand over the items in person. I believe the scheme is more efficient now than it was a few years ago, but nothing is perfect: workload issues and job changes can all have an effect on the process.

I often record my coin finds with The Fitzwilliam Museum Corpus of Early Medieval Coin Finds (EMC). This is a project that records single coins dating between AD 410 and 1180, so any Saxon, Norman and early Plantagenet coins that you are lucky enough to find can be recorded simply and extremely efficiently by sending an email to Dr Martin Allen at the museum's Department of Coins and Medals with any information you can provide about when and where they were found. You will receive details back such as the age of the coin, type and class, even where the coin was minted and who the moneyer was! I find the process extremely helpful and Dr Allen is kind enough to respond with the information very quickly. However, to help with the identification of the coin you must attach high quality images, any blurred pictures will hamper his efforts in trying to understand the finer details. It is worth noting that EMC does not give valuations of finds, it is purely for identification purposes and to assist the project.

Contact details for Dr Allen are here:

Dr Martin Allen
Department of Coins and Medals
Fitzwilliam Museum,
Cambridge,
CB12RB
email: mra25@cam.ac.uk

One particular example of how good the EMC is for recording and identifying coins stands out in my mind. Late in 2018, I was fortunate enough to unearth a silver hammered coin, of a type I had not seen before. The rest of the detecting session had me wondering what it could be. I knew by the style and appearance that it was an early coin, had fixed my mind on it being late Saxon and looked forward to getting home to research it a little more. After a little time researching, I finally discovered what it was: a King Stephen and Queen Matilda penny; one which turned out to be extremely

rare. However, I was completely stumped when it came to any more information about it, but I sent a quick email to Dr Allen and within a day he had replied with really interesting information. He revealed where it was made (York) and who minted it, and even provided information on the dies that were used to strike the coin. These were unmatched, which made the coin even more desirable. The only work I had to do was take a decent quality image of the front and back, a brief description of the county and area it was found and that was it: a really efficient service. So you can see the reasoning behind recording your coins and artefacts: the information you get back is just as exciting as actually finding the items in the first place.

If you require further information on the Portable Antiquities Scheme you can simply log onto their website at www.finds.org.uk. Here you will find other links you may find useful, including information on recording finds in Scotland, Northern Ireland and the Isle of Man.

A Rare King Stephen and Matilda penny found by the author.

CHAPTER EIGHT

How to preserve and care for your discoveries

ALTHOUGH, THERE IS a great temptation to clean an object, if you feel, for any reason, that it might have significant historical or archaeological value, then it is better to leave it as found until you get expert advice.

Some coins and artefacts will be stable and need little or no cleaning. Copper coins for example that show detail will benefit from a dry environment for twenty-four hours or so, then with a little care can be lightly brushed to remove any dry mud and soil. This will reveal a nice green patina, a date and the king or queen's head for you to enjoy studying at your leisure. That is how I deal with my Victorian pennies and halfpennies for example, and you can spend many an hour checking the dates and condition of them and maybe sharing the coins with the landowner. Equally there will be coins that you unearth that will simply be corroded beyond all recognition; no amount of care and cleaning will help. I have seen people try their best to reveal something from these worn-out discs, scrubbing with wire wool, soaking them in Coca Cola, you name it, they have tried it; the only result will be a scoured mess. 'I'm sure I can see a head of a king, yep, there is a cross or shield on here, look!' In reality people are just seeing marks made by trying to over-clean a worn coin. I can fully understand why people do it, because I did it myself, but it really isn't worth the time and effort. (Please note that the advice I am giving here is for everyday finds that you are likely to make, objects of little or no monetary value and of no significant historical interest.)

Before we go any further I must stress that my own methods of cleaning and preserving some of the coins and artefacts I find may

A silver Roman denarius before careful cleaning.

differ greatly from others, so it is always the best policy to get a second opinion with regards to your own finds.

As mentioned above, a dry environment is crucial for your finds; a plastic container with the objects placed within is a good idea. Find a safe place to store them for twenty-four hours or more, but not near a radiator or fire as this will dry the items out too quickly and can cause damage. I find room temperature is best. Once the finds that you have made are dry, you can use a soft-bristled toothbrush to gently remove any soil that may be stuck to the item; take care around any decoration or detail and just brush over the muddy deposits to remove them, this will then reveal the coin or artefact for you to examine and enjoy its true beauty.

For many years many of the copper and bronze coins I unearthed would be soaked in olive oil. However, I have stopped using this method as the oil can have a slightly negative effect on the objects if soaked for too long (although many people still use the oil method). I learnt from a friend a few years ago that the organic acids contained in olive oil which act as the cleaning agents, vary in concentration depending on the pressings of the olives. Extra virgin oil, being the first pressing, has the lowest concentration of acids so it is not really a good agent for cleaning coins. So if you do decide on olive oil then just use the plain type; if soaking coins over a period of time to try and lift any debris from the surface then it is wise to change the oil on a regular basis. A number of individuals I know use mineral oil; being non-organic it seems to achieve better results when using it

Silver Roman denarius revealed after careful cleaning. It shows the head of Hadrian who was Emperor from 117 CE to 138 CE.

on coins and artefacts. Whichever cleaning method you choose, take care: stability is the key, along with common sense.

Eventually, at some point while wandering around with your detector, you will unearth a silver coin; some will be as bright and shiny as the day they were made, others will be heavily tarnished. There are a few methods of cleaning the tarnish off that I would avoid, and some that I would use with a little caution, so let's have a look at these in a little more detail.

For many centuries, silver coins have been minted, and as you may imagine they can come up in a variety of conditions – and I have had my fair share of bad ones. Over time, some coins may develop a build-up of crud, especially Roman denarius coins. A concretion develops on both sides that can be very difficult to remove, I find the best method for actually breaking down the build-up is to place the coin in a plastic tub and pour a little lemon juice over it, just enough to cover the coin and leave for a few minutes; you can repeat this process a few times until the detail of the coin is revealed. Use a piece of kitchen cloth to gently wipe after the soak and you will notice how black the tarnish/concretion is. Eventually the coin will return to its former detail and you will be amazed how you have turned this featureless disc into something stunning. However, not everyone agrees with this method, some will advise against it, but it works for me and has done so for over three decades. I've saved these little works of art and made them attractive and desirable to view and I'm comfortable with that. Some people that use this method go a stage further and use a teaspoon of bicarbonate of soda with the lemon juice, a reaction takes place and a little fizz which seems to help break the crud and tarnish down; once completed a rinse under the tap will reveal a decent enough coin if you haven't gone overboard with the method. Again, common sense plays a part and like everything cleaning wise, should always be done with caution: too much could ruin a good coin.

Many years ago, I came across a gentleman who had a rather crude method of cleaning his silver coins, especially hammered silvers: when he uncovered a heavily tarnished coin, he placed it in between two pieces of tinfoil, dampened it under a tap and rubbed gently for a minute or so. This revealed the coin free from the tarnish but heavily scratched. It was especially apparent when placed under a magnifying glass; yes a lot more detail was visible, but the coin was now damaged and completely worthless. I would strongly advise anyone against using this method, although people still do it to this day.

As mentioned earlier, drying a coin or artefact and making sure it is stable goes a long way to preserving it, but what about cleaning around any detail a find may have? Well I often use a wooden toothpick to gently remove any stubborn and dried mud that may have embedded into areas of design; a buckle for example may be highly ornate, but a lot may be hidden under lines of dried soil. I take the sharpness slightly off the end of the toothpick then gently work the mud away, sometimes working around the detail for some time; eventually, with the dirt gone, a rather nice and collectable artefact will be revealed for you to display. To stabilise the item even further, I sometimes apply a layer of microcrystalline wax, which is really good for preserving an artefact or coin over a long period. This wax is available in many outlets in the UK; just search for 'cleaning products for detecting finds' for example.

Other popular cleaning accessories that detectorists often use are electrolysis, ultrasonic and barrelling machines. These have been used for a number of decades and prove to be very effective in cleaning coins. Electrolysis is particularly effective, but it is a method that needs serious attention to ensure it works correctly, as it involves using a power source, wires with crocodile clips, and the correct voltage. It works by passing a small electrical charge from the power source through the metal item while it is submerged in an electrolyte solution. Always seek advice before using this method as it can seriously strip any stubborn deposits down very quickly and cause damage to the item you are cleaning. Some detectorists make their own electrolysis machines and electrolyte solutions, but once again you really need to be comfortable with what you are doing before you become successful at using it.

Barrelling machines were extremely popular back in the 1980s and 1990s and people still use them for cleaning their coins and artefacts. However, it is strongly advised that individuals only clean solid copper alloy objects in the machine as it can be a very harsh method; anything with enamel or that is plated will be damaged. Never use a barrelling machine for silver or gold, as the items will be completely ruined.

Finally, ultrasonic coin cleaners are also popular with some detectorists. These accessories can vary in price so it is worth doing your homework and read reviews before deciding which is the best to purchase. The machine works by producing high frequency sound vibrations; these exert a very strong action that dislodges dirt and encrustation from objects without any abrasion.

With any coins and artefacts, common sense comes into play; please do not fall into the over-cleaning trap and if an item is stable and appealing then leave it that way. I cannot think of anything worse than someone trying to improve an already desirable object and I shudder to think how many important and valuable coins have been ruined by people over-cleaning them.

Remember, this chapter is based on my advice only, so please also seek professional information if you do decide to clean any of your detector finds.

CHAPTER NINE

Enjoying the Hobby and Outdoors: The Most Important Aspect of the Pastime

WE HAVE COVERED information regarding rules, regulations, codes of conduct and the law; now is the time to really open up about just how unique, rewarding, magical, captivating, addictive (and sometimes frustrating) this hobby can be. I will try not to get carried away and do my best to keep a lid on my enthusiasm!

Detectorists enjoying the winter sunshine.

A medieval coin seeing light for the first time in many centuries.

To be able to search the fields that our ancestors once walked, worked and rested in is truly magical; unearthing the everyday items they carried with them or had attached to their belts; their shoes and clothing; items worn around their necks or nestled on their fingers. It not only transports you back in time but gives you a real sense of the distant past – looking at a featureless field in the middle of nowhere you would be hard pressed to imagine the amazing items that may be hiding just a few inches under the soil – yet with a little time and effort and a metal detector, we can bring these back to light for everyone to see.

Just imagine the scene; it is early autumn, the field that you have been patiently waiting to search has finally been turned over by the farmer's machinery and is now flat enough for you to walk on, the sun is warm, but the gentle breeze takes the edge off making it too uncomfortable. As you scan the surface with your detector, you notice tiny flecks of cockle shell, fragments of old pottery mixed in with stone and flint. Your eyes are firmly fixed on the surface of the field, gliding your search coil over the small lumps of soil, waiting for that signal, that tone to come through the headphones that could indicate something of interest. The last three targets uncovered were a twisted piece of unidentifiable metal and two shotgun cartridge caps, but you are sure something of interest will turn up soon: you have that gut instinct and all the signs are indicating you are in the right area. By lunchtime, more cartridge ends, strangely shaped white lead and a button have filled your finds pouch but that hasn't dampened your enthusiasm.

You head back to the car for a welcome bite to eat and a drink and realise it is a stunning day; smelling that fresh air it feels good to be alive, a world away from being stuck in an office or hidden away in a murky factory with little or no daylight to see. A hare runs across the hedgerow in front of you and stops to assess any danger, for the first time today you notice the birdsong and oh that sky, wide open and vast. You could stand and watch it forever and you feel alive and stress-free as you walk back to the area you were searching that morning. Moving to the side of your earlier footprints you start work again, slowly, methodically scanning the area you are convinced will produce something of note. More unidentifiable items of copper, lead and modern rubbish follow the same pattern as before and slight frustration grows; do you stick or twist? The next ten minutes are spent thinking about moving to the lower end of the fields; you check the settings on the detector just to make sure everything is working correctly and then it happens: a signal is registered, a slight two-way bleep that has you down on your knees with your spade. A shovel-full of soil containing the target is out on the surface, you run the search coil over the soil to home-in on the target and there it is. You get the first glimpse of a round grey-looking disc, half-covered in dirt, which you gently extract. Scanning the item with your eyes, you roll it over in the palm of your hand to get a better identification. You make out a cross on one side and from your pocket you take a small bottle of water and gently pour it on the coin. Before your eyes, the mud washes off to reveal a silver medieval penny from the reign of Edward I and right there you are transported back in time. Indeed it is the closest thing to time travel you can get. 'Who was the last person to hold this little piece of history and where did they live? What did they look like? What job did they do?' you wonder, as you sit gazing at the object. As you get back to your feet and place the coin safely in your finds' box you then set about trying to find another; by close of play you are tired but very content with your day's searching and as you drive home you reflect on the day and just how good it was. But the magic doesn't stop there; unlike other hobbies, metal detecting continues to captivate the imagination, as the very next day you are busily trying to identify and uncover more information about the item you found: where the coin was minted, what class type is the coin, and of course the condition of it. The value is immaterial; it's the fact that you have unearthed a time capsule from the medieval period: that is the true worth.

Days after, you find yourself showing the coin to friends and family; proud of your discovery you start to give history lessons to anyone that will listen. Some will only be interested in its value and dismiss the small disc because it is not dripping with gold, but others will see the true beauty and be keen to hold something that is several centuries old and quite possibly the oldest object they have ever handled. Then as the coming weekend nears, the whole adventure starts again; regardless of the weather, you will be systematically searching to bring the past to light.

Now you could be fooled into thinking that the anecdote above is based around my own experiences, but this could quite easily happen to any of the many thousand avid detectorists who share the understanding, the passion and the way of life. As I have said many, many times, this hobby is the nearest thing to time travel you will get, as a pastime nothing comes close to it.

So as you can see, the picture of metal detecting is starting to take shape, but let us also look at what else the hobby offers.

HEALTH AND WELL-BEING

I cannot stress enough just how big a part health and well-being plays in this hobby, People from all walks of life, who may suffer from anxiety, depression and other mental health issues benefit greatly from being outdoors; participating in any hobby is of great help, but detecting, in my opinion seems to really have the edge. Not only does it free the mind from burdens, work-related issues or the daily grind of life, I find it actually conditions the mind to concentrate on a completely different path.

On a personal level, metal detecting has helped me overcome some dark periods throughout the years, and I speak to so many people who use the hobby to escape the ups and downs that life can bring; the distraction helps ease the troubles of everyday issues and gives people real focus. The improvement on mental health is also coupled with physical health and people who wouldn't necessarily get outdoors that much find themselves walking many miles while participating in the hobby. Concentration levels are heightened and of course that wonderful fresh air adds to the benefits. I have lost count of how many people I know that have taken up the hobby after retirement, both men and women, who now go out detecting

Many happy faces at a detecting rally.

most weekends and weekdays when possible. Not only that but they spark up new friendships with like-minded people, especially if they attend rallies or organised digs. So as you can see metal detecting plays a much bigger role than simply unearthing coins and artefacts.

RAISING MONEY FOR CHARITIES

Some time ago, I attended a rally in Lincolnshire, where a group of detectorists are invited onto farmland for a paid day's detecting. The proceeds raised usually go to the landowner and dig organiser for allowing access to the land and organising the event. If carried out in a professional manner these events can be a very enjoyable occasion and offer people the chance to detect on land they would not normally have access to, and also the opportunity to find some interesting items on the day. However, this particular rally was a little different, in that it was raising money for the county's air ambulance; it was not just a one-off: those involved had been doing this periodically for a number of years, raising several thousands of pounds in the process. What struck a chord with me after speaking with a few of the detectorists on the day, was that it was not an isolated event: a similar theme was happening in other counties, all raising money for different charities and really making a difference. Every week, detectorists up and down the country are raising vital funds for not only the air ambulance but for veterans' charities, the homeless, cancer research – the list goes on. The combined money raised over the years must easily run into the millions. There's no drama, no fuss: the organisers just get on with it quietly and efficiently, so I would like to take this opportunity to congratulate them all for the amazing work they do.

TIDYING UP THE LANDSCAPE

One area of work not often spoken about is the removal of scrap from fields and beaches that detectorists do every day of the week. You only have to take a peek inside someone's finds' pouch and you will see an array of twisted metals, lead, iron fragments, pieces of farm machinery or aluminium. Some items are potentially harmful for the environment: I've even seen hypodermic needles and other dangerous implements come up from the earth. You will be surprised what is buried out there and once again without the determination and hard work of the detectorist, it would all be hidden and left to rot; some in the food growing fields too. Only recently I ventured out onto some coastal beaches near to where I live, to hopefully enjoy a few hours searching on the wet sand. By lunchtime I was trudging back to the car with a carrier bag full of ring pulls, melted disposable barbeques and a hefty load of beer cans. I lost count how many other pieces of scrap I removed, so let it be said, metal detecting works for the community!

Recycling waste can be a real problem.

THE FIELD

Situated on the outskirts of a small village this particular patch of land isn't eye catching, in fact you could quite easily walk passed it without showing a blind bit of notice to it. Bordered by a wonky hedgerow on one side and a typical country lane on the other it awkwardly works its way down to a point and finishes at a 'past its sell by date' fence that could do with some serious attention.

For 20 years or so we have detected this 'awkward patch', mainly because it's one of the only fields available to us late on in the season, the farmer usually leaves it in stubble to rot down or in a flattened state once the sugar beet comes off. A few finds have come off it over the years, enough to make us venture back on it each season, with the hope of a decent hammered or a medieval artefact or two, some years we come off with smiles on our faces and some years we don't, pretty much like it is with most fields.

However, something is different about this one, a sadness usually sets in whilst I roam up and down its acreage, and that's because not that long ago we used to detect as a small team of 3, sometimes 4, enjoying a little banter before and after the detecting session, sharing a little food in the hedgerow and generally looking at each other's finds whilst trampling off at dusk. A long drive home was usually filled with jokes, laughter and a few raised eye-brows when the snoring starts if someone nodded off! Now it's just the two of us, myself and Garry, who trail the soil, up and down, in the areas of the field where we used to detect alongside the other two, Shaun and Mo, both passed away within a year or so of each other, they loved detecting, and shared similar values to myself and Gaz.

It's a strange, surreal feeling, caught up in a situation you think would never happen, both their voices in my head, showing me coins, telling which area they found a brooch or where the pottery scatter showed up more than in other areas, maybe it's me, and my nature, but I can't help thinking of them every time I'm on that field, as we gathered as a small group on here more than any of the others, the noise of their coils thrashing through stubble or their distant voice shouting me over to show a coin or two. It's sad, so incredibly sad to think they can no longer see what I see, to enjoy the experience of digging signal after signal, in the hope of that special find, just being out there trying their hardest at the hobby they loved. so the moral of this story is, make the most of the time you have with your friends whilst out in the fields, cherish every minute of it, if you don't find much then so what, at least you are out there, with good company enjoying the fresh air, Good detecting partners are hard to find, I had 3, now I'm down to one…. And I will never take it for granted.

LOCATING LOST ITEMS

It would be foolish to finish this chapter without mentioning another vital area of work that detectorists offer: a free service to help find lost items. These could be important objects, such as car keys, or sentimental and personal items such as engagement and wedding rings. Every week I receive a call from some poor soul who has either lost a ring while gardening or their car keys have slipped from their pocket while out walking their dogs. If they are local, I generally arrange to meet with them to show me where they think the items could be, or if they are further afield I do my best in locating a detectorist in that particular area via social media to see if they will offer their assistance. Most of the time we can reunite the lost items with their owners, which can be quite an emotional experience for them, especially if the item has been lost for some time.

One particular example stands out in my mind and I shall never forget it. A few years ago, I received a call from a man who was a distraught because he had lost a very precious wedding ring, which his wife had especially made for him. However, this wasn't a normal call from someone who had lost it in the garden or walking his dog, it had been lost on a building site while he was moving tonnes of earth with machinery. At around lunchtime he unfortunately spilt diesel on his hands, which in turn loosened the ring from his finger; it was a few hours before he noticed it had gone and he feared the worst. Clutching at straws he managed to locate my name and address via an Internet search and within the hour I arrived on site armed with my detecting gear. At first glance I didn't hold much hope of finding it, the whole site was a muddy mess of machinery,

A beautiful gold ring recovered for its owner.

earth and standing water, but I tried my best to keep his spirits up. One thing I've learnt over the years with detecting is never give up, but after thirty minutes of searching the area where he had been working we had nothing to show for it: not a whisper of anything metallic came to light. At this stage the gentleman became quite upset as he told me his wife had the ring specially made while she was abroad; not only was it unique but very expensive. Time was running out and rain had set in, 'the clock was ticking' so to speak. Then, in desperation I scanned the detector over an area of long marsh grass that was growing in large patches nearby, a fair distance away from where the man had said he may have lost the ring. After a few minutes, the detector registered a slight blip and as I forced the coil down further into the long grass the signal became stronger. As I reached down into the marshy earth I glimpsed the unmistakable colour of gold; getting to my feet again I opened my hand to reveal the most amazing, heavy and ornate gold ring I have ever seen. The two images of the ring show its sheer size and beauty; we both stood rooted to the spot for some time, both with beaming smiles. In another five minutes we would have simply called it a day and the ring, due to the development work going on, would have been lost forever. The sheer emotion on the gentleman's face was humbling – he actually shed a tear or two, it meant that much to him. I never saw him again, nor did I catch his name, but the memory of finding that very special ring will stay with me forever.

So as you can see, there is so much more to metal detecting than simply searching for coins and artefacts. I've tried to capture the magic of the hobby within this one chapter, but in reality it would take many, many books to do that. If I had to sum it up then it would always be these words, 'my hobby, our hobby, is full of spirit, imagination, determination, discovery, magic and adventure'. I cannot find its equal, simply because there isn't one.

A stunning gold medieval coin found by the author.

CHAPTER TEN

Amazing Treasure Finds from the UK

MOST PEOPLE WHO take up the hobby of metal detecting dream of finding that once-in-a-lifetime discovery: a hoard, an artefact, something of special historic significance – a reward for the countless hours spent out in the fields and on the beaches. In stark reality this seldom happens and only a small percentage of people actually strike it lucky. Of course many treasure items are unearthed each year which build a detailed picture of the historical landscape for us all to enjoy, but not many detectorists out there really hit the big one, regardless of how many hours they spend searching. However, that doesn't stop us all trying our hardest every week!

In this final chapter we look at some of the amazing finds that detectorists have uncovered over the years in the UK; it goes without saying that without their luck, perseverance and skill these treasures would still be buried in the ground and quite possibly be under threat from agricultural damage from machinery etc.

A clod shot of a Celtic silver coin.

THE HOXNE HOARD

In 1992, Eric Lawes unearthed the largest hoard of Roman treasures ever found here in the UK. At the time he was searching for a hammer which his friend had lost earlier that week. Eric was using his new metal detector, a retirement present, on the day that turned out to be one he would never forget. The hoard comprised of 14,865 Roman gold, silver and bronze coins, and around 200 items of silver tableware and gold jewellery.

At first Eric uncovered gold coins and silver spoons, but after finding many of these decided to call the landowners The next day a team of archaeologists arrived to excavate the area and the entire hoard was brought to light.. The hoard had been buried in a wooden chest sometime in the late Roman times, a period of

Item of treasure from the Hoxne Hoard.

great upheaval due to the collapse of Roman rule, and one can only speculate why this hoard was buried. Raids, theft, wealthy owners under threat from their enemies, who knows, but we can all use our imagination here to come up with countless stories about this wonderful discovery. Incidentally the hammer was also found and was passed over to the British Museum to form part of the overall collection. In 1993, the Treasure Trove Committee valued the hoard at £1.75 million, which was paid to Eric and the farmer Peter Whatling. However, I think you will all agree that the historical value of this wonderful find is priceless.

THE STAFFORDSHIRE HOARD.

In 2009, a detectorist named Terry Herbert found the largest hoard of Anglo-Saxon gold and silver ever discovered. He was searching on land in Staffordshire when he came across some 3,500 items, scattered over a large area and thought to date to the seventh century. The hoard itself is virtually all military in its nature and the workmanship of the items are of the highest detail, making this hoard, like the Hoxne discovery very much of national importance. The landowner agreed for the area to be excavated and the story of this most important discovery began to take shape. Later down the line, when the hoard was displayed at the Birmingham Museum and Art Gallery it attracted more than 40,000 people, who like me and many other detectorists must have been captivated by the magic of this amazing find. The hoard was finally valued at £3.285 million and split between both Terry and the landowner, but once again the true value lies in its historical significance. So much information has come to light since this discovery, and research projects have been set up off the back of it – once again this is all down to the original work of the detectorist.

Amazing detail on a Saxon artefact from the Staffordshire Hoard.

Treasure items from the Staffordshire Hoard.

A stunning image of the Middleham Jewel.

THE MIDDLEHAM JEWEL

The Middleham jewel is a fifteenth-century gold pendant, which was unearthed by a detectorist in 1985. On each side of the pendant there is an engraved religious scene and one side has a large sapphire set into it. The amazing artefact was found near Middleham Castle and it is thought that the pendant is linked to Richard III because it is a high status item and only the very rich and wealthy would own something of this nature. The finder, Mr Ted Seaton, found it on a bridle path; at a Treasure Trove inquest it was actually declared as lost or abandoned and sold at auction for £1.6 million. After an export licence was refused, money was then raised and it was acquired by the Yorkshire Museum in 1992 for £2.5 million. The jewel is now on permanent display in the museum where many people have enjoyed its beauty over the years. Imagine if Ted hadn't ventured out onto that bridle path, the pendant and its beauty might still be hidden.

THE GROUVILLE HOARD

Detectorists Reg Mead and Richard Miles unearthed a huge hoard of late Iron Age and early Roman coins which they reported in 2012. The hoard, unearthed in the parish of Grouville on Jersey, was estimated to be made up of over 70,000 coins. An earlier, smaller hoard was found in the area in the early 1980s and the two detectorists eventually found the main part, which weighed in at an incredible 750 kg. It is thought to have belonged to a local Celtic

The sheer size of the Grouville Hoard.

Caption

The Grouville Hoard excavated.

tribe that may have been fleeing Julius Caesar's armies sometime around 50 BC. The determination shown by the two detectorists is another example of the work we do: by painstakingly searching for years on end for a hoard they knew was somewhere nearby (the landowner only granted them permission to search the land once a year, just after harvest time), but their perseverance certainly paid off! The value of the hoard is unclear as the Treasure Trove process may no longer be applicable in Jersey, but in a historical context this hoard is hugely important. Interestingly, gold torques, silver bracelets, and other items were mixed in with the mass of coins.

The Frome Hoard being carefully excavated.

THE FROME HOARD

In 2010, metal detectorist Dave Crisp unearthed a huge hoard of over 52,000 Roman coins. The coins were debased silver and bronze and the hoard is one of the largest ever uncovered in the UK. The date range of the coins is between AD 253 and 305. Searching

The Frome Roman coin Hoard.

in a field near Frome where he had previously found Roman coins, Dave received what he described as a strange and funny signal; digging down he found the top of a pot and realised it was in fact an intact hoard. Once again we can only imagine why the coins may have been buried. Whether they were hidden by the owner to be retrieved at a later date or whether they were a votive offering to the gods, we will never know. The hoard was eventually valued at £320,250 and was acquired by the Museum of Somerset in 2011. It was reported that in twenty-two years of detecting it was the most significant find that Dave had ever made.

THE SILVERDALE HOARD

In 2011, detectorist Darren Webster unearthed a truly amazing hoard of Viking treasure, consisting of more than 200 pieces of silver jewellery and coins, near Silverdale in Lancashire. The items were buried in a lead container and at first Darren thought he had found a sheet of lead, but as he raised it up a little, silver objects were literally spilling out of it and at this stage he knew he had discovered something rather special. The authorities were contacted and the hoard was handed over to the British Museum for observation and analysis. Within the hoard were twenty-seven coins, finger rings, arm rings, brooches and ingots, but what was particularly interesting to the museum while studying the hoard, was that one of the silver coins had been stamped with the name of a previously unknown Viking ruler. In 2011, the discovery was deemed as treasure and was later valued at £110,000. It was eventually purchased by the Lancashire County Museum and has been on display in various locations throughout the north-west.

Darren's find turned out to be the fourth largest Viking discovery in the UK, and does make you wonder just how many more are lying out there waiting to be unearthed. Do not rule out another detectorist locating another similar hoard soon.

The list of vital, historical finds made by detectorists goes on and on and the details described are just a tiny example of the essential but sometimes overlooked work of the passionate detectorist. I sometimes shake my head in disbelief when these amazing treasures are brought to light and showcased through the media but with little or no mention of the detectorist who uncovered it. Thankfully, with the emergence of social media we all get the opportunity to give metal detectorists the credit they deserve by sharing their success stories. Many times over the years when news breaks about a particular treasure discovery I briefly separate the detectorist from the find, and imagine their morning being just like any other. Breakfast, detecting gear ready, car radio on, driving to their field of choice for that day, completely oblivious to what is about to happen. A brief chat to the landowner, a little banter maybe about finding treasure and then it's off to the field in question. A stretch of the legs, back and arms when getting out of the car and straight into the boot where a scene of chaos unfolds while frantically searching for the headphones, followed by the one glove that's missing. Finally

The Silverdale Viking hoard on display.

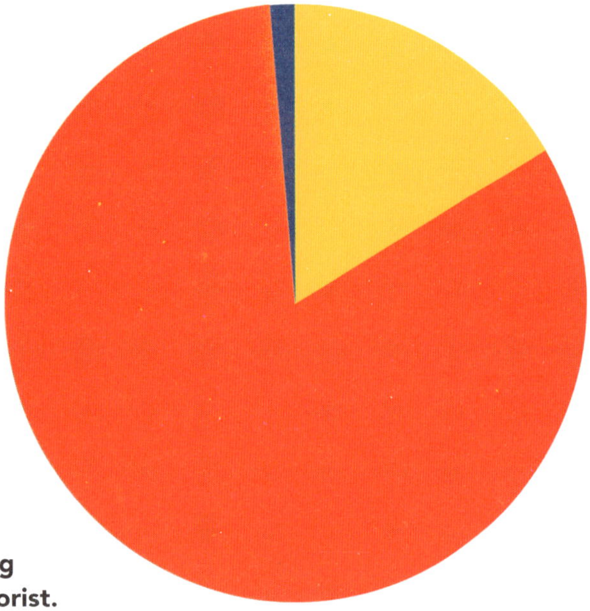

A pie chart clearly showing the sterling work of the detectorist.
(Image courtesy of Steve Lancaster).

everything is in place and it's off to the top corner of the land near to the village. Detector on, searching commences and it's pretty much the same as the last time: broken pieces of metal, cow tubes, ring pulls, the odd tin can and shotgun shell, until just before lunch a target is located that needs investigating. The hole is dug and in all probability it's another piece of farm machinery, yet something in the corner of the earth catches the eye; just a glimpse of something out of the ordinary. That shape, that colour; the mind is saying 'no chance, it's a coloured piece of discarded metal, nothing to see here', until the hand reaches down and gently touches the item, the fingers run down the side until it meets the earth where the rest is buried. Some soil is removed to reveal a partly buried gold Iron Age torque, but still the brain is working against it, you haven't found this, it's got to be a wind up, how can this possibly happen to me . . .

The reality is that this could happen to anyone of us: regardless of how slim the chances are, each time we venture out detecting we simply do not know what lies ahead of us.

Finally, I cannot finish this book without mentioning once again that we are all ambassadors of the hobby, we all have a part to play in promoting the vital work we do when bringing history back to life. Never miss an opportunity to talk to people who may simply show an interest in what you are doing in the fields or down on the beach. They will be curious to know what you are finding, interested in the machine you are using and may wish to learn how they could get involved in the pastime themselves. Many of you may have found your way into the hobby from this very route.

Pick the day, grab your detector and get outdoors, for this is where your own adventure begins. Happy searching!